The Grumpy Dwarfgoby:

Discovering the Red Sea's Hidden Gem

Could This Tiny Fish Change How We See the Ocean's Future?

Kai Silverton

Table of Contents

Introduction: The Tiny Fish with a Big Attitude 5

Chapter 1: Unveiling the Red Sea's Hidden Treasure 8

Chapter 2: Meet the Grumpy Dwarfgoby (Sueviota aethon) 13

Chapter 3: The Science Behind the Discovery 18

Chapter 4: Living on the Edge – The Grumpy Dwarfgoby's Habitat 24

Chapter 5: Threats to Coral Reefs and Marine Biodiversity . 30

Chapter 6: The Role of Tiny Fish in the Coral Reef Ecosystem 36

Chapter 7: Conservation and What We Can Do 41

Chapter 8: More Wonders of the Red Sea 47

Chapter 9: A Closer Look – Comparing the Grumpy Dwarfgoby with Other Dwarfgoby Species 53

Chapter 10: The Future of Marine Exploration 58

Chapter 11: The Call to Action 63

    Acknowledgments 67

Appendices 69

Disclaimer

The information presented in this book is for educational and informational purposes only. While the author has made every effort to provide accurate, up-to-date, and well-researched content, this book should not be considered as professional advice in the fields of marine biology, ecology, or conservation. Readers are encouraged to consult experts and official resources for specific guidance or clarification on matters relating to marine science and conservation efforts. The author and publisher disclaim any liability or responsibility for any direct or indirect damages, losses, or consequences resulting from the use of information contained in this book. The views expressed within are those of the author and do not necessarily reflect the views of scientific organizations or entities mentioned herein.

Copyright © 2024 [Kai Silverton]. All Rights Reserved.

No part of this book may be reproduced, distributed, or transmitted in any form or by any means, including photocopying, recording, or other electronic or mechanical methods, without the prior written permission of the publisher, except in the case of brief quotations embodied in critical reviews and certain other non-commercial uses permitted by copyright law.

Introduction: The Tiny Fish with a Big Attitude

In the heart of the Red Sea, one of the world's most vibrant and biologically diverse marine ecosystems, a remarkable discovery was made—a tiny fish, no longer than a few centimeters, with an unmistakable scowl that earned it the name "grumpy dwarfgoby." Despite its small size, this new species has sparked significant excitement among marine biologists and environmentalists alike, symbolizing both the diversity and fragility of marine life in a rapidly changing world. The journey to uncovering the *Sueviota aethon*, or grumpy dwarfgoby, was as much an adventure as it was a scientific breakthrough. During a routine dive to explore the uncharted biodiversity of the coral reefs, a team of researchers, including ecologist Viktor Nunes Peinemann, spotted a curious, bright red fish nestled deep within crevices of red algae-covered corals. At first glance, it appeared to be a known species of goby, but something about its grumpy expression, defined by its sharp canines and stocky build, prompted further investigation. Back in the lab, under close analysis, the researchers realized they had stumbled upon a completely new species. The grumpy dwarfgoby had gone unnoticed for so long due to its exceptional camouflage, blending seamlessly into the reddish coral reef environments. This fish, though diminutive, represents a massive scientific find—proof that even in well-studied ecosystems like the Red Sea, nature still holds surprises for those who dare to look closely. But what makes the grumpy dwarfgoby truly fascinating isn't just its discovery; it's what this tiny fish represents in the broader context of marine biodiversity. In a world where ecosystems are increasingly under threat from climate change, pollution, and human activity, the discovery of a new species reminds us of the delicate balance within coral reefs and other marine environments. These ecosystems are not just homes to charismatic megafauna like sharks and sea turtles but also to minute, intricate organisms like the grumpy dwarfgoby that play essential roles in maintaining ecological balance. The Red Sea, where the grumpy dwarfgoby resides, is a unique marine habitat teeming with life. It hosts hundreds of species that exist nowhere else in the world, many of which are critical to the health of coral reefs. The grumpy dwarfgoby, with its specific adaptations, such as its small size and reliance on coral crevices for protection, is a perfect example of how evolution has shaped species to fit into the narrowest ecological niches. Coral reefs, often referred to as the "rainforests of the sea," depend on these tiny fish and other organisms to sustain their incredibly diverse ecosystems. What's more, the discovery of the grumpy dwarfgoby comes at a time when coral reefs worldwide are in peril. Rising ocean temperatures, acidification, and pollution have contributed to the bleaching of coral reefs and the decline of their inhabitants. In many ways, the grumpy dwarfgoby's "grumpy" appearance could serve as a metaphor for the distress marine ecosystems are facing today. This discovery is not just about adding a new species to the record books; it's a call to action to protect the fragile environments that are home to countless creatures like the grumpy dwarfgoby.

The grumpy dwarfgoby, though tiny and seemingly insignificant, has become a symbol of the mysteries that remain hidden beneath the ocean's surface. It represents the importance of continuing to explore, study, and protect the oceans and the life within them. Every new discovery serves as a reminder that while we have mapped much of the earth's surface, the depths of our oceans remain vastly unexplored, with untold species awaiting discovery. As we delve deeper into this book, you'll come to understand not just the specifics of the grumpy dwarfgoby, but also the broader story of marine biodiversity, the challenges facing our oceans, and the delicate balance that holds these ecosystems together. In the following chapters, we will explore the process of discovery, the unique traits of this fascinating fish, and why protecting species like the grumpy dwarfgoby matters more than ever in today's world.

This tiny fish, with its big attitude, has more to teach us than we might expect.

Chapter 1: Unveiling the Red Sea's Hidden Treasure

The Red Sea, a narrow body of water separating the Arabian Peninsula from Africa, has long been a destination of fascination for marine biologists, ecologists, and adventurers alike. Renowned for its vibrant coral reefs, crystal-clear waters, and rich biodiversity, it serves as a sanctuary for countless marine species. Among the most recent discoveries is the *grumpy dwarfgoby*—a tiny fish that encapsulates the incredible diversity of life in the region. This chapter explores the uniqueness of the Red Sea ecosystem, the coral reefs that support the grumpy dwarfgoby, and the exciting journey that led scientists to uncover this new species.

An Overview of the Red Sea's Unique Ecosystem

The Red Sea is one of the most extraordinary marine environments on Earth. Stretching approximately 2,250 kilometers from the Suez Canal in Egypt to the Bab el Mandeb Strait near Yemen and Djibouti, it is home to a wealth of biodiversity. What makes the Red Sea particularly remarkable is its unique combination of geographical features and environmental conditions, which support a wide array of life forms.

One of the key factors contributing to the Red Sea's ecological richness is its **temperature stability and salinity levels**. Unlike other coral reef systems, which are often susceptible to seasonal fluctuations, the Red Sea remains relatively warm throughout the year, with temperatures ranging between 20°C and 30°C (68°F and 86°F). This stability, combined with its relatively high salinity (due to low freshwater influx), has allowed the development of some of the most **resilient coral species** in the world.

The Red Sea's coral reefs are considered **"super reefs"** because they have shown an unusual resilience to rising ocean temperatures and acidification, conditions that are threatening coral systems globally. This resilience may be one reason why species like the grumpy dwarfgoby thrive here. Scientists have noted that while coral bleaching is occurring in many parts of the world, the reefs of the northern Red Sea have remained largely unaffected. This resilience is critical in the face of climate change and makes the Red Sea a hotspot for marine biodiversity research.

Beyond coral reefs, the Red Sea boasts **mangroves, seagrass beds, and extensive coastal lagoons**. These habitats create a complex network that supports a rich variety of marine life—from large species like dolphins, sharks, and manta rays to smaller, less noticeable creatures like the grumpy dwarfgoby. This ecosystem diversity is a key feature that has long drawn the attention of scientists, with the understanding that each element plays an essential role in the health of the marine environment.

The Coral Reef Biodiversity that Supports the Grumpy Dwarfgoby

The coral reefs of the Red Sea are among the most biologically diverse ecosystems on the planet, second only to rainforests in terms of species richness. These reefs serve as the primary habitat for the grumpy dwarfgoby, along with thousands of other marine species. With over **300 species of coral** and **1,200 species of fish**, the Red Sea's reefs are an intricate and vibrant mosaic of life, where every organism—from the smallest goby to the largest predator—has a role to play in maintaining ecological balance.

The **grumpy dwarfgoby** thrives in the hidden nooks of these coral structures, particularly among the **red coralline algae** that cover many of the reefs. These algae not only provide camouflage for the goby but also serve as a rich feeding ground for small invertebrates, which are a key part of the goby's diet. The coral reefs provide not just physical protection from predators but also a stable environment where the goby can reproduce and flourish.

One of the reasons coral reefs are such hotbeds of biodiversity is that they offer a wide range of microhabitats. Each species of coral creates unique structures—branches, plates, and mounds—that different species of fish, crustaceans, and invertebrates use for shelter and food. The grumpy dwarfgoby is adapted to live in the smallest crevices, taking advantage of spaces that larger fish cannot reach. In many ways, the goby's tiny size is its greatest asset, allowing it to occupy ecological niches that other species cannot.

The **symbiotic relationships** between coral species and marine organisms are crucial to the health of the reef. Coral polyps, the tiny animals that make up coral reefs, engage in a symbiotic relationship with **zooxanthellae**, algae that live within their tissues. This relationship allows corals to photosynthesize and build the vast reef structures that support the entire ecosystem. Species like the grumpy dwarfgoby rely on the stability and productivity of these coral systems to thrive.

However, the health of these reefs is under constant threat from human activities, including **pollution, overfishing, and tourism**. Despite its resilience, the Red Sea is not immune to the global forces of climate change, which makes the preservation of such ecosystems crucial not only for the survival of species like the grumpy dwarfgoby but for the countless others that depend on coral reefs worldwide.

How Scientists First Stumbled Upon This Species

The discovery of the grumpy dwarfgoby is a story of curiosity and perseverance. It occurred during an expedition led by **Viktor Nunes Peinemann** and a team of marine biologists who were exploring the northern Red Sea's lesser-known coral reefs. Their initial goal was to catalog various fish species living in the region, but they soon encountered something unexpected.

During a routine dive near one of the Red Sea's many coral outcrops, Peinemann and his team observed a tiny, reddish fish with a rather distinctive appearance. At first glance, they believed it to be a species of goby they had seen before, but upon closer inspection, they noticed subtle yet significant differences. The fish had prominent **canine-like teeth**, which were unusual for gobies of its size, and its unique coloration, perfectly camouflaged against the red algae, was striking.

Back on the research vessel, the team captured detailed images and took samples for genetic analysis. What they discovered shocked them: this was not just another goby—it was an entirely new species. Named *Sueviota aethon*, the grumpy dwarfgoby was officially recognized and classified as a new species, cementing its place in marine biology records.

The discovery of the grumpy dwarfgoby highlights the importance of continued exploration in even the most studied environments. Despite centuries of research, the oceans remain largely unexplored, and new species like the grumpy dwarfgoby remind us of the mysteries still waiting to be uncovered. For the scientists involved, it was a thrilling moment, one that emphasized the need for further conservation efforts in the Red Sea.

As the following chapters will show, the discovery of the grumpy dwarfgoby is not just an isolated event—it is part of a much larger story about the ongoing exploration of the oceans and the urgent need to protect the ecosystems that harbor such incredible biodiversity.

Chapter 2: Meet the Grumpy Dwarfgoby (Sueviota aethon)

The *Sueviota aethon*, affectionately known as the grumpy dwarfgoby, may be small, but it packs a big personality. This chapter will delve into the species itself, focusing on its physical characteristics, habitat preferences, and its vital role within the ecosystem of the Red Sea. While its "grumpy" look might attract attention, its behavior and adaptability in the fragile coral reef environment make it a fascinating creature to study.

Detailed Description of Its Physical Features and Distinctive "Grumpy" Look

The grumpy dwarfgoby is a visually striking fish, even among its colorful coral reef neighbors. Measuring only about **2 centimeters** in length (roughly the size of a small pea), this tiny fish stands out because of its vivid reddish coloration, which matches its preferred habitat—**red coralline algae-covered reefs**. Its bright color serves a dual purpose: both camouflage within its environment and a potential signal to other species of its presence.

But the feature that has earned this goby its memorable nickname is its **"grumpy" expression**. This is a combination of its **slightly protruding lower jaw**, where two small but prominent **canine-like teeth** are visible, and its somewhat furrowed "brow," created by the shape of its eyes and face. These physical traits give the grumpy dwarfgoby a permanent frown, a feature that is particularly endearing to observers but serves no aggressive purpose. In fact, despite its grumpy demeanor, this goby is more likely to hide than confront larger predators or even similarly sized competitors.

Additionally, the grumpy dwarfgoby has a stocky body that is slightly compressed, with large **pectoral fins** that help it maneuver effortlessly within the tiny crevices of coral reefs. Its fins also aid in short bursts of swimming to catch prey or evade danger. The **caudal fin**, or tail, is rounded and compact, ideal for short-range movements rather than long-distance swimming.

A unique feature of gobies is their **lack of a swim bladder**, which is common in many other fish species to help maintain buoyancy. Without this organ, the grumpy dwarfgoby is better suited to spending its life near the reef's surface, where it can rest on corals and algae-covered rocks without floating away.

The Dwarfgoby's Habitat: Red Algae-Covered Coral Reefs

The habitat of the grumpy dwarfgoby is a critical part of its survival and success as a species. It thrives in the

red algae-covered coral reefs of the Red Sea, an environment perfectly suited to its needs. The reefs themselves are dynamic ecosystems teeming with life, but the **red coralline algae** in particular play a pivotal role in the dwarfgoby's daily activities.

Red coralline algae, named for their calcified, reddish exterior, are an integral part of coral reef environments. These algae not only contribute to the structural integrity of the reef but also create numerous crevices and hiding places, which are essential for small creatures like the grumpy dwarfgoby. The fish uses these crevices for shelter from predators, protection from ocean currents, and as vantage points from which to ambush prey.

The Red Sea, known for its high salinity and warm temperatures, creates a unique environment where species like the grumpy dwarfgoby can flourish. The abundance of small crevices and shelters in red algae-covered coral reefs makes this habitat a prime location for the goby, which is often spotted darting in and out of algae beds in search of food or to avoid predators. Unlike larger fish, the dwarfgoby's size allows it to access areas of the reef that would be impossible for bigger species, further safeguarding it from danger.

The coral reefs in the Red Sea provide more than just shelter for the grumpy dwarfgoby—they are also rich in the invertebrates and microorganisms that make up the goby's diet. As reefs flourish, so does the life within and around them, creating an intricate food web of which the dwarfgoby is a key part.

Behavior, Diet, and Ecological Role Within the Reef

Despite its tiny size, the grumpy dwarfgoby plays an important ecological role in maintaining the balance of its coral reef ecosystem. Like other gobies, it is **territorial**, often defending its tiny space within the reef from intruders. However, due to its small size and lack of physical prowess, its defense strategy is more about **evasion and quick bursts of movement** than direct confrontation. Its agility allows it to dart into the smallest cracks in the coral, disappearing from sight when threatened.

The grumpy dwarfgoby's **diet** primarily consists of **tiny crustaceans, plankton, and small invertebrates** that are abundant in the coral reef environment. Its feeding behavior involves hovering near the coral or algae beds, waiting for small prey to drift by in the water column. It then uses quick, darting movements to snatch its prey before retreating to its hiding place. This **ambush strategy** is typical of many goby species and is particularly effective for a fish that lives in such a densely populated and competitive environment.

Ecologically, the grumpy dwarfgoby contributes to the health of the coral reef by participating in the **food web** that sustains the reef ecosystem. By preying on small invertebrates, it helps to control their populations, which, if left unchecked, could damage the coral structures themselves. In turn, the goby serves as prey for larger fish and marine predators, creating a balance that maintains the biodiversity and stability of the reef.

Furthermore, the presence of species like the grumpy dwarfgoby in coral reefs is often an indicator of the overall health of the ecosystem. Small fish species are typically more sensitive to environmental changes, and their population density can provide clues about the condition of the reef. In this way, the grumpy dwarfgoby can be seen as a **biological marker**, signaling whether a coral reef is thriving or in decline.

The grumpy dwarfgoby, with its distinctive appearance and important ecological role, is a perfect example of how even the smallest creatures can have a significant impact on the health of their environment. Its reliance on the red algae-covered coral reefs of the Red Sea highlights the interconnectedness of marine species and the importance of conserving delicate ecosystems like coral reefs. As we continue to explore the nuances of this species, the next chapter will take a deeper dive into the importance of coral reefs and why protecting them is crucial not just for the grumpy dwarfgoby, but for marine biodiversity as a whole.

Chapter 3: The Science Behind the Discovery

Unveiling the grumpy dwarfgoby was not just a stroke of luck, but the result of careful scientific investigation and fieldwork. This chapter dives into the rigorous process behind identifying a new species, examining how the *Sueviota aethon* was recognized, what makes it stand out compared to similar species, and the firsthand experiences of the researchers who were instrumental in its discovery.

The Scientific Process of Identifying a New Species

Identifying a new species involves a series of methodical steps that scientists must follow to ensure their findings are credible and accurate. The discovery of the grumpy dwarfgoby followed the traditional pathway of field observation, specimen collection, morphological analysis, and genetic testing.

Field Observation and Collection:
The first step in the discovery process occurred in the coral reefs of the Red Sea, where the small, reddish fish was spotted in its natural habitat. Researchers, such as Viktor Nunes Peinemann and his team, were conducting biodiversity surveys of the region. During a routine dive, they observed an unfamiliar fish in the red algae-covered reefs, its distinctive appearance catching

their attention. Field observations included noting the fish's habitat, behavior, and interaction with the reef ecosystem, all while taking careful photographs and video documentation.

To confirm that this was indeed an unknown species, several individuals were **collected** using specialized techniques. Scientists ensured minimal disturbance to the reef while retrieving specimens for further analysis. Collecting live specimens is crucial, as it allows for detailed examinations of physical traits and genetic material.

Morphological Analysis:
Once the specimens were brought to a laboratory, the next step was to examine their **morphological features**—essentially, the physical traits that differentiate this goby from others. This analysis typically involves using **microscopy and detailed measurements** to compare the new species with known species in the same family.

In the case of the grumpy dwarfgoby, scientists examined key features such as its **body size, fin shape, and distinctive dental structure**. The two small canine-like teeth that gave the species its "grumpy" appearance were a standout characteristic. These physical traits were meticulously cataloged and compared against other species of the *Sueviota* genus and similar gobies.

Genetic Testing:
In modern taxonomy, morphological analysis alone is not enough to classify a new species. Genetic analysis,

particularly **DNA barcoding**, plays a critical role in confirming whether an organism belongs to a previously undiscovered species. In the case of the grumpy dwarfgoby, samples were taken for **DNA sequencing**, where scientists compared its genetic material to that of other known gobies. The DNA sequencing revealed significant genetic differences, confirming that the *Sueviota aethon* was indeed a new species, genetically distinct from its closest relatives.

Once the morphological and genetic data were gathered, the research team published their findings in peer-reviewed journals, a critical step in getting the species officially recognized by the scientific community. The species was then added to global taxonomic databases, where it was given its formal name.

Comparative Analysis with Similar Species and Why This Goby Stands Out

The discovery of the grumpy dwarfgoby is particularly exciting because, at first glance, it shares many traits with other species of gobies found in coral reefs. However, a closer look reveals some unique features that set this species apart.

Similar Species:
The grumpy dwarfgoby belongs to the genus *Sueviota*, a group of tiny gobies typically found in coral reefs across the Indo-Pacific region. Other species within this genus share similar habitats, body sizes, and feeding behaviors. For example, species like *Sueviota atrinasa*

and *Sueviota pyrios* are known for their ability to blend into their surroundings and live in the nooks and crannies of coral reefs.

However, despite these similarities, the grumpy dwarfgoby stands out for several reasons:

1. **Distinctive "Grumpy" Look:** The most visually striking feature is its face, which appears to frown due to the prominent canine-like teeth and the configuration of its jaw. While many gobies have slight variations in facial structure, the combination of its lower jaw protrusion and dental formation gives this species its distinctive, perpetually "grumpy" expression.
2. **Coloration and Habitat Preference:** Unlike other species in the genus *Sueviota*, the grumpy dwarfgoby is specialized to live in red coralline algae, which directly influences its reddish pigmentation. Most gobies exhibit duller colors to blend with sandy or rocky environments, but the *Sueviota aethon*'s reddish hue is perfectly matched to the red algae that cover the reefs, offering it an additional layer of camouflage.
3. **Unique Genetic Signature:** Genetic analysis further confirmed that the grumpy dwarfgoby has enough **genetic divergence** from its closest relatives to be classified as a distinct species. This divergence suggests that the species may have evolved in relative isolation, adapting specifically to its niche environment in the Red Sea, a region known for its unique

species due to its geographical and environmental characteristics.

Insights from the Researchers Who Made the Discovery

The excitement surrounding the discovery of the grumpy dwarfgoby is shared by the scientists who made it. **Viktor Nunes Peinemann**, the lead researcher on the project, described the moment of discovery as both exhilarating and unexpected. According to Peinemann, the team was initially focused on surveying known species in the Red Sea, but the goby's unusual appearance immediately stood out.

In interviews and publications, Peinemann emphasized the importance of **exploration in marine biology**, noting that even in well-studied regions like the Red Sea, new discoveries are still possible. He also highlighted the value of **collaboration**—their team included marine biologists, taxonomists, and geneticists, all working together to confirm the new species. Peinemann was quick to acknowledge that discoveries like these are the result of both **luck and meticulous planning**. The initial observation of the fish's unique features led them down the path of further investigation, ultimately confirming that they had found something new.

One of the most rewarding aspects for the team was seeing their research contribute to the **understanding of coral reef biodiversity** in the Red Sea. For Peinemann, the discovery underscores how much we still have to learn about the oceans, and why conservation efforts are critical in protecting these environments from degradation. In particular, he stressed the need for **protecting coral reefs**, which serve as the foundation for so many marine species, including the newly discovered grumpy dwarfgoby.

As we move forward in the book, the discovery of the grumpy dwarfgoby becomes part of a larger conversation about the ongoing exploration of the world's oceans, the challenges facing coral reef ecosystems, and the importance of conservation in preserving marine biodiversity for future generations.

Chapter 4: Living on the Edge – The Grumpy Dwarfgoby's Habitat

The Red Sea, where the grumpy dwarfgoby thrives, presents a unique and dynamic environment, shaped by its geography and climate. This chapter will explore how the specific conditions of the Red Sea contribute to the survival of this species, the vital role that coral reefs play in maintaining marine biodiversity, and the various adaptations the grumpy dwarfgoby has developed to navigate life in this fragile ecosystem.

How the Red Sea's Unique Conditions Support This Species

The Red Sea is unlike many other marine environments due to its **high salinity, elevated water temperatures**, and relatively isolated location. Stretching from the **Gulf of Suez** in the north to the **Bab el Mandeb Strait** in the south, this semi-enclosed body of water experiences minimal freshwater inflows from rivers, leading to higher than average salinity levels—around **40 parts per thousand**, compared to the **35 ppt** found in most oceans. This salty water creates a challenging environment for marine species, but those that do live here, like the grumpy dwarfgoby, are highly specialized to survive in such conditions.

The **warm temperatures** of the Red Sea, which range from **22°C (71°F)** in the northern parts to over

30°C (86°F) in the south, have also shaped the species that live there. These warm waters enable the growth of **coral reefs**, which provide shelter and food for countless organisms, including the grumpy dwarfgoby. The coral reefs of the Red Sea are particularly unique because they are more **thermally tolerant** than many others, which has allowed them to survive while other reefs worldwide have been devastated by **coral bleaching** caused by rising ocean temperatures.

For the grumpy dwarfgoby, these environmental factors create both opportunities and challenges. The warm, salty waters of the Red Sea have led to the development of **highly specialized niches** where only certain species, such as the grumpy dwarfgoby, can thrive. The abundance of **red coralline algae** in these reefs provides both shelter and a hunting ground for the dwarfgoby, helping it evade predators while securing its food supply. However, these conditions also mean that the goby is particularly vulnerable to any environmental changes, such as **climate change** or **pollution**, which could disrupt the delicate balance of the reef ecosystem.

The Importance of Coral Reefs to Marine Life

Coral reefs are often referred to as the "rainforests of the sea" due to the incredible amount of biodiversity they support. Although they cover less than **1%** of the ocean floor, they are home to **25%** of all marine species. The Red Sea's coral reefs, in particular, are crucial to maintaining the biodiversity of the region, providing habitat, food, and breeding grounds for a multitude of species, including the grumpy dwarfgoby.

The structure of coral reefs creates a variety of **microhabitats** that serve different species' needs. **Branching corals** offer hiding places for smaller fish like the dwarfgoby, while **massive corals** provide stable surfaces for algae to grow, which is essential for herbivorous species. The grumpy dwarfgoby's habitat, nestled in the crevices of red coralline algae, highlights the importance of these microhabitats.

Additionally, coral reefs act as **nurseries for juvenile fish**, protect coastlines from erosion, and support local economies through **fisheries and tourism**. For example, coral reefs in the Red Sea attract thousands of divers annually, contributing significantly to the region's economy. **Egypt's coral reef tourism alone** brings in **over $7 billion** annually

Reef Builders

, a figure that underscores the importance of conserving these habitats.

However, these ecosystems are under constant threat from human activities such as **overfishing, coastal development**, and **climate change**. The **2015 coral bleaching event**, for example, affected many parts of the world, although the Red Sea's reefs fared better due to their natural resilience. Still, the grumpy dwarfgoby and other reef species remain vulnerable, highlighting the critical need for **conservation efforts** to protect these invaluable ecosystems.

Adaptations of the Grumpy Dwarfgoby for Survival in a Fragile Ecosystem

Living in the dynamic and sometimes harsh conditions of the Red Sea, the grumpy dwarfgoby has developed several adaptations that enable it to survive in this fragile ecosystem.

1. **Size and Behavior Adaptations:**
   At just **2 centimeters in length**, the grumpy dwarfgoby's diminutive size is one of its most important adaptations. Its small size allows it to hide in the tiny crevices formed by red coralline algae and coral structures. This behavior of **staying close to shelter** at all times is key to avoiding predators, which are abundant in the Red Sea's reefs. By using short, darting movements, the goby can quickly disappear into its hiding spots whenever a threat appears.
2. **Camouflage:**
   The goby's **reddish coloration** is perfectly suited to its environment. As it spends most of its life nestled among **red algae**, this camouflage helps it blend into its surroundings, making it harder for predators to spot. This color matching also aids in **ambush feeding**, where the goby waits for unsuspecting prey to drift by before lunging from its hiding place.
3. **Dietary Flexibility:**
   The grumpy dwarfgoby has a **generalist diet**, feeding on **small plankton, invertebrates**, and algae that are abundant in coral reef ecosystems. This dietary flexibility allows it to

make the most of the resources available in its environment, ensuring it can survive even when certain prey species are scarce. Its ability to switch between **active hunting** and **grazing** on algae provides it with a level of resilience in a habitat where food sources can fluctuate based on environmental conditions.

4. **Reproductive Strategy:**
   Like many small reef fish, the grumpy dwarfgoby is likely to have a **high reproductive rate**, producing large numbers of eggs during the breeding season. This ensures that even though many of its offspring will fall prey to larger fish, enough will survive to sustain the population. This reproductive strategy is essential in an ecosystem where the odds are often stacked against smaller species.

5. **Resilience to High Salinity and Temperature:**
   One of the most remarkable adaptations of the grumpy dwarfgoby is its **tolerance to high salinity and temperature**. The Red Sea's conditions, particularly in its northern regions, are far more extreme than many other marine environments. Over time, the goby has adapted to thrive in these salty, warm waters. However, this specialization also makes it highly vulnerable to any drastic changes in temperature or salinity, which could push the species beyond its limits of tolerance.

The grumpy dwarfgoby, despite its small size, has evolved a remarkable set of adaptations that allow it to thrive in the unique and sometimes challenging conditions of the Red Sea. The coral reefs it calls home are a critical part of this survival, offering both shelter and sustenance. However, as coral reefs face increasing threats from climate change and human activity, the future of this species, like many others, hangs in the balance. As we look ahead, the conservation of coral reefs will be essential to ensuring that the grumpy dwarfgoby and its ecosystem can continue to thrive.

Chapter 5: Threats to Coral Reefs and Marine Biodiversity

Coral reefs like those found in the Red Sea, home to the grumpy dwarfgoby, are facing increasing threats from various environmental and human-induced factors. This chapter explores how **climate change, coral bleaching, ocean acidification, habitat destruction**, and human activities such as **overfishing and pollution** are jeopardizing these fragile ecosystems and the species that depend on them.

The Impact of Climate Change on Coral Reefs and Species Like the Grumpy Dwarfgoby

Climate change is arguably the most significant threat to coral reefs and the marine species they support. Rising global temperatures are affecting the oceans in several ways, leading to conditions that disrupt the delicate balance of coral ecosystems.

1. **Increased Sea Surface Temperatures**: One of the most direct effects of climate change is the warming of the oceans. Coral reefs are highly sensitive to temperature changes; even a rise of **1-2°C** above normal sea temperatures can cause coral bleaching. The Red Sea, though naturally warmer, is not immune to these changes. Prolonged exposure to **higher**

temperatures can stress the corals, leading them to expel the symbiotic algae that give them color and provide essential nutrients through photosynthesis.

Without these algae, the corals not only lose their vibrant colors but also their ability to sustain marine life, including the grumpy dwarfgoby. Reefs that experience frequent or severe bleaching events can die off completely, transforming biodiverse areas into barren, underwater deserts. This loss of habitat would directly affect species like the grumpy dwarfgoby, which rely on coral for shelter, breeding grounds, and food sources.

2. **Shifts in Ocean Currents and Weather Patterns**:
   Climate change is also affecting ocean currents, rainfall, and storm patterns. For example, **stronger storms** can cause physical damage to coral reefs, breaking apart delicate structures. Changes in **current patterns** can affect nutrient flow and disrupt the migration of plankton, which forms a critical part of the grumpy dwarfgoby's diet. The delicate balance of reef ecosystems, already finely tuned to local conditions, could be upended by these broader changes, putting additional stress on both corals and marine species.

Coral Bleaching, Ocean Acidification, and Habitat Destruction

While rising temperatures are the most immediate threat to coral reefs, other factors are contributing to their decline as well, with **coral bleaching, ocean acidification**, and **habitat destruction** being at the forefront.

1. **Coral Bleaching**:
   As mentioned earlier, coral bleaching occurs when corals lose their symbiotic algae due to stress, primarily caused by **heat stress**. In recent years, bleaching events have become more frequent and severe, leaving many reefs struggling to recover. Some areas of the **Great Barrier Reef**, for example, have lost over **50% of their coral cover** due to repeated bleaching events

   Popular Science

   . While the Red Sea's corals have shown a remarkable ability to withstand higher temperatures, they are not invincible. Repeated bleaching events could erode their resilience, leaving the grumpy dwarfgoby without the shelter and resources it needs to survive.

2. **Ocean Acidification**:
   Another byproduct of climate change is **ocean acidification**, caused by the oceans absorbing excess carbon dioxide from the atmosphere. As CO2 dissolves in seawater, it forms **carbonic acid**, which lowers the pH of the water. This

increase in acidity makes it harder for corals to form the **calcium carbonate skeletons** that form the structure of reefs. Weakening coral structures could lead to the collapse of entire reef systems, further endangering species like the grumpy dwarfgoby that depend on these reefs for survival.

Acidification also affects other marine species, such as **shellfish** and **plankton**, which play key roles in the food web. As the availability of these organisms diminishes, the entire ecosystem could suffer. The grumpy dwarfgoby, which feeds on plankton and other small invertebrates, would be directly impacted by such changes, facing food shortages as its prey populations dwindle.

3. **Habitat Destruction**:
   Beyond climate-related threats, human activities continue to cause significant damage to coral reefs. **Coastal development**, dredging, and **destructive fishing practices** such as **blast fishing** can physically damage or destroy reef systems. In the Red Sea, rapid tourism development has increased pressure on coral reefs, with **diving** and **snorkeling activities** sometimes contributing to coral breakage or anchor damage.

   Habitat destruction disrupts the complex web of life that exists within coral ecosystems, displacing species and fragmenting populations. For small species like the grumpy dwarfgoby, which rely on specific coral formations for protection, habitat destruction can be catastrophic.

How Overfishing and Pollution Endanger Marine Species

In addition to environmental pressures, human activities such as **overfishing** and **pollution** are further endangering coral reefs and the species that depend on them.

1. **Overfishing**:
   Overfishing has a cascading effect on coral reef ecosystems. Large fish, which are often the target of fishing efforts, play a critical role in maintaining the balance of reef systems by keeping populations of **herbivorous fish** and **invertebrates** in check. When these predator species are removed, herbivorous fish populations can surge, leading to **overgrazing** of algae, which can smother coral reefs.

   Alternatively, in some areas, overfishing of herbivorous species such as **parrotfish** and **surgeonfish** has allowed algae to overgrow, outcompeting corals for space and sunlight. This shift can lead to the collapse of the reef ecosystem, depriving the grumpy dwarfgoby of its shelter and food sources.

2. **Pollution**:
   Pollution from various sources—such as **plastic waste, oil spills**, and **chemical runoff** from

agriculture—has a profound impact on coral reefs. **Microplastics**, for example, can be ingested by marine organisms, including plankton, disrupting the food chain. Chemical pollutants, particularly **fertilizers** and **pesticides**, can lead to **eutrophication**, where nutrient overloads cause massive blooms of algae that suffocate coral reefs.

Sedimentation from coastal development can also cloud the water, blocking sunlight from reaching corals and disrupting the photosynthesis process that is crucial for their survival. For the grumpy dwarfgoby, this can mean reduced visibility, making it harder to hunt for prey and increasing its vulnerability to predators.

Coral reefs, including those in the Red Sea, are at a critical juncture. Threats from climate change, habitat destruction, overfishing, and pollution are converging to create a challenging environment for marine life. As we continue to witness the degradation of coral reefs globally, it becomes increasingly important to prioritize **conservation efforts**. For species like the grumpy dwarfgoby, the loss of coral ecosystems would mean not only the loss of a habitat but also a significant threat to their survival.

Chapter 6: The Role of Tiny Fish in the Coral Reef Ecosystem

Despite their size, small fish like the grumpy dwarfgoby play an essential role in the health and stability of coral reef ecosystems. This chapter delves into why these smaller species matter, how they fit into the web of interdependent organisms within the reef, and why preserving biodiversity, particularly of tiny species, is crucial for maintaining the ecological balance of coral reefs.

Why Small Species Like the Grumpy Dwarfgoby Matter

1. **Ecosystem Engineers in Miniature**: Small fish species such as the grumpy dwarfgoby may not seem as impressive as larger predators, but they are vital to the functioning of coral reefs. Their **presence maintains microhabitats** within the reef, providing both shelter and food sources for other small marine organisms. For instance, the dwarfgoby often inhabits nooks and crevices in **algae-covered coral reefs**, and in doing so, keeps these areas from being overgrown by algae that could otherwise suffocate corals.

Additionally, their ability to find and use small, hidden spaces creates **safe havens** for larvae of other fish species, which need protection from predators. In this way, the grumpy dwarfgoby plays a role in the **nursery functions** of coral reefs—where young fish can grow until they are large enough to survive in the open ocean.

2. **Food Web Contributions**:
   Though small, these fish are essential components of the **reef food web**. They feed on **plankton** and other tiny invertebrates, helping regulate the populations of these organisms. Without small species like the grumpy dwarfgoby to control plankton levels, the ecosystem could experience overpopulation of these species, disrupting the delicate balance that coral reefs depend on.

   In turn, the grumpy dwarfgoby serves as **prey for larger species**, such as **groupers** and **moray eels**. By forming an integral part of the food chain, these small fish help sustain the populations of larger predators, which in turn keep the populations of other species in check. Their presence ensures that **nutrients** circulate throughout the ecosystem, fostering biodiversity and keeping the reef healthy.

The Interdependence of Coral Reef Organisms

Coral reefs are some of the most diverse ecosystems on Earth, home to thousands of species that interact in intricate, interdependent ways. The grumpy dwarfgoby is just one piece of this vast ecological puzzle, but its role is critical in maintaining balance.

1. **Symbiotic Relationships**:
   Coral reefs are sustained by **symbiosis**, where different species depend on each other for survival. For example, coral itself has a symbiotic relationship with **zooxanthellae**, the algae that live within its tissues and provide energy through photosynthesis. The grumpy dwarfgoby benefits from this system, as healthy coral supports a thriving environment full of resources for the fish.

   Beyond direct interaction with coral, species like the grumpy dwarfgoby may form **mutualistic relationships** with other reef inhabitants, such as **cleaner fish** that help remove parasites or scavenger species that consume organic material, keeping the environment clean.

2. **The Domino Effect of Species Loss**:
   The coral reef ecosystem relies on the continued existence of even its smallest residents. **Losing small fish species** like the grumpy dwarfgoby can have a **cascading effect** throughout the ecosystem. Without them, the microhabitats they maintain could become overrun by harmful algae, the food web would suffer, and larger

predators would lose a food source, potentially leading to their decline as well.

These complex connections mean that each organism contributes to the overall resilience of the reef. Coral reefs thrive on diversity, where every organism, no matter how small, plays a role in its survival. A **decline in biodiversity** can make the entire system more vulnerable to environmental stressors such as climate change, overfishing, and pollution.

The Ecological Importance of Maintaining Biodiversity

1. **Biodiversity as a Buffer**:
   The presence of a wide variety of species ensures that coral reefs are more resilient to changes in environmental conditions. Biodiversity provides an **ecological buffer**—if one species is lost, others can often fill its role, allowing the ecosystem to continue functioning. However, with the loss of multiple species, the ecosystem's ability to adapt diminishes, making it more likely to collapse under stress.

   In the case of small species like the grumpy dwarfgoby, their **role in nutrient cycling**, **predation**, and **microhabitat maintenance** is irreplaceable by larger or less-specialized species. As coral reefs face increasing pressures from climate change and human activities, the importance of preserving these small, often overlooked species becomes even more crucial.

2. **Keystone Species Impact**:
   While small in size, species like the grumpy dwarfgoby could be considered **keystone species** in their specific niches. A **keystone species** is one whose presence or absence significantly influences the structure of the ecosystem. Although typically thought of as large predators, some small fish exert a disproportionately large impact on the health and function of their environments.

   By maintaining the balance of microhabitats, regulating prey populations, and contributing to the reef's complex food web, the grumpy dwarfgoby plays a **keystone role** within the coral reef ecosystem. Protecting small species like this is essential to the long-term sustainability of coral reefs.

In conclusion, while often overlooked, small species like the grumpy dwarfgoby are crucial to the function and health of coral reef ecosystems. Their contributions to **nutrient cycling**, **habitat maintenance**, and **biodiversity** illustrate the importance of conserving even the tiniest organisms. Ensuring that these fish continue to thrive not only supports the reefs they call home but also contributes to the overall health of our oceans.

Chapter 7: Conservation and What We Can Do

Coral reefs are essential for marine biodiversity, supporting ecosystems that house countless species, including the newly discovered grumpy dwarfgoby. Unfortunately, these vital ecosystems are under threat from a variety of sources. In this chapter, we explore **current conservation efforts** focused on the Red Sea and similar habitats, provide **practical ways for readers to get involved** in marine conservation, and offer **actionable steps** for individuals to reduce their environmental impact.

Current Efforts to Protect the Red Sea and Its Inhabitants

Several international and local initiatives are actively working to protect the Red Sea, its coral reefs, and the species that rely on them, including the grumpy dwarfgoby. These efforts focus on **preservation, restoration, and sustainable management** of marine environments.

1. **Marine Protected Areas (MPAs)**:
   One of the most significant strategies to conserve marine biodiversity is the establishment of **Marine Protected Areas** (MPAs). These designated zones restrict harmful activities like fishing, pollution, and coastal development. MPAs in the Red Sea aim to safeguard coral reefs from overexploitation while allowing ecosystems to regenerate naturally. Notably, areas like the **Farasan Islands MPA** in Saudi Arabia and the **Ras Mohammed National Park** in Egypt protect extensive coral reefs and their inhabitants from human interference. The creation of more MPAs around the Red Sea, including in regions where the grumpy dwarfgoby has been found, is a critical step in preserving marine life.
2. **Coral Restoration Projects**:
   Coral restoration is another important initiative in the Red Sea. Efforts to **replant damaged coral** or **grow new coral fragments** are underway in many regions. These projects focus on replenishing coral that has been lost due to bleaching or physical damage from human activities. Innovative methods, such as **coral gardening** or **artificial reef structures**, are used to encourage coral growth, allowing ecosystems to recover faster. By ensuring the reefs are healthy, these initiatives indirectly support the survival of species like the grumpy dwarfgoby.
3. **Sustainable Tourism and Development**:
   The Red Sea is a popular destination for diving and tourism, but without proper management,

these activities can harm delicate ecosystems. To address this, local governments and international organizations have begun promoting **sustainable tourism practices** that prioritize the protection of coral reefs. Programs that teach tourists about **reef-friendly behaviors**—such as avoiding touching corals or wearing **reef-safe sunscreens**—are crucial for minimizing human impact.

Governments are also pushing for **sustainable coastal development**, ensuring that resorts, ports, and other infrastructure are designed to minimize their ecological footprint. Sustainable development helps to protect habitats like the one in which the grumpy dwarfgoby thrives.

How Readers Can Contribute to Marine Conservation Efforts

While large-scale conservation projects are vital, individual actions can also have a meaningful impact. Here's how readers can get involved in protecting marine ecosystems like those in the Red Sea:

1. **Support Marine Conservation Organizations**:
   One of the simplest ways to contribute is by **donating** to or **volunteering** with organizations dedicated to marine conservation. Groups like **The Coral Restoration Foundation, Ocean Conservancy**, and **The Marine Conservation Institute** lead projects globally to protect coral reefs, restore damaged

ecosystems, and raise awareness about marine issues.

Many of these organizations offer ways for the public to engage directly, whether through **volunteer programs**, **beach cleanups**, or **educational campaigns**. By getting involved with these efforts, readers can make a direct contribution to preserving the ocean's biodiversity.

2. **Adopt-a-Coral or Marine Animal**:
   Some organizations offer programs where individuals can **adopt a coral**, marine animal, or even a specific portion of a reef. These symbolic adoptions help fund ongoing conservation efforts and raise awareness. Programs such as **"Adopt A Coral"** allow participants to sponsor coral fragments that will be replanted, directly contributing to coral reef restoration.
3. **Participate in Citizen Science**:
   Citizen science projects allow everyday individuals to assist researchers in monitoring marine ecosystems. For example, divers can report coral bleaching or unusual fish sightings, helping scientists track the health of reefs and detect changes in biodiversity. Platforms like **iNaturalist** and **Reef Check** invite participants to share observations from dives or beach walks, providing valuable data to researchers studying marine ecosystems like the one that houses the grumpy dwarfgoby.

Practical Steps for Individuals to Reduce Their Environmental Footprint

Reducing our collective environmental footprint is crucial to alleviating the pressures on marine ecosystems. Here are some simple yet effective ways that individuals can lessen their impact:

1. **Reduce Plastic Use**:
   Plastic waste, particularly **single-use plastics**, is one of the most pervasive pollutants in the ocean. To reduce plastic pollution, individuals can:
   - **Use reusable bags, bottles, and containers** instead of single-use plastics.
   - **Avoid products with excessive packaging** and choose alternatives to plastic whenever possible.
   - Participate in **beach or river cleanups** to help remove plastic waste before it reaches the ocean.
2. **Choose Sustainable Seafood**:
   Overfishing is one of the primary threats to marine biodiversity, so opting for **sustainable seafood** can make a big difference. Organizations like the **Marine Stewardship Council (MSC)** provide certifications for seafood products that are sustainably sourced. By purchasing certified products, individuals can help reduce overfishing and protect marine species.
3. **Reduce Carbon Emissions**:
   Climate change is one of the biggest threats to

coral reefs and marine life, so reducing your carbon footprint is a key way to support marine conservation. Practical steps include:
- **Using energy-efficient appliances** and reducing energy consumption.
- **Opting for public transportation**, biking, or walking instead of driving.
- Supporting policies that encourage **renewable energy** and **carbon reduction** initiatives.

4. **Choose Eco-friendly Tourism**:
For those planning trips to coral reef destinations, it's essential to choose **sustainable tourism operators**. Look for companies that prioritize reef protection, limit visitor numbers, and educate tourists on how to enjoy the reef without causing harm. Tourists can also help by supporting **reef-safe sunscreens** that don't contain harmful chemicals like **oxybenzone**, which contribute to coral bleaching.

Chapter 8: More Wonders of the Red Sea

The Red Sea is a biodiversity hotspot, renowned for its rich coral reefs and unique marine life. In this chapter, we explore other fascinating species in this region, examine why the Red Sea's ecosystem is one of the most ecologically valuable on Earth, and discuss the future of marine exploration in this unique environment.

Other Fascinating Species of the Red Sea's Coral Reefs

The Red Sea is home to a variety of species that are found nowhere else in the world. Its coral reefs shelter over 1,200 species of fish, 10% of which are endemic to this region. Some of the most captivating species include:

1. **Napoleon Wrasse (*Cheilinus undulatus*)**: This large, colorful fish is a popular sight for divers. Known for its distinctive forehead bump

and vibrant green, blue, and yellow hues, the Napoleon wrasse plays a vital role in maintaining the health of coral reefs by controlling populations of harmful invertebrates such as crown-of-thorns starfish.

2. **Red Sea Bannerfish (*Heniochus intermedius*)**: The Red Sea bannerfish is striking with its long, flowing dorsal fin and bold black, white, and yellow stripes. It forms schools near coral reefs, feeding primarily on plankton. Its presence contributes to the balance of the marine food web, keeping plankton populations in check.

3. **Masked Butterflyfish (*Chaetodon semilarvatus*)**: One of the most iconic fish of the Red Sea, the masked butterflyfish is easily recognized by its bright yellow color and distinct blue "mask" around its eyes. These fish are often seen in pairs and are crucial to coral health as they feed on polyps, controlling excessive coral growth and maintaining the reef structure.

4. **Red Sea Clownfish (*Amphiprion bicinctus*)**: Like its famous relatives in the Indo-Pacific, the Red Sea clownfish has a symbiotic relationship with sea anemones. These fish find shelter in the stinging tentacles of the anemones while helping keep them clean by removing debris and parasites.

5. **Giant Moray Eel (*Gymnothorax javanicus*)**: The giant moray eel, reaching lengths of up to 3 meters, is one of the top predators in the Red Sea's coral ecosystems. With its powerful jaws and keen sense of smell, it hunts at night,

playing an important role in regulating fish populations.

These species illustrate the incredible biodiversity within the Red Sea's coral reefs. Each organism, from small invertebrates to top predators, contributes to a complex web of interdependence, ensuring the resilience and survival of this fragile ecosystem.

Why the Red Sea's Biodiversity is Ecologically Valuable

The Red Sea's biodiversity is a result of its unique geographical and environmental characteristics. Unlike many other coral ecosystems, the Red Sea experiences relatively high salinity levels and water temperatures, fostering a distinct set of species that are adapted to these conditions.

1. Endemism and Evolutionary Significance: The Red Sea's isolated location, bordered by deserts and limited in freshwater inflow, has led to high levels of endemism. This means that many species found in this region have evolved in response to its specific environmental challenges, making them unique to the Red Sea. Endemic species like the Red Sea clownfish and certain corals demonstrate this region's evolutionary importance, acting as living laboratories for studying adaptation and speciation under extreme conditions.

2. Resilience to Climate Change: Coral reefs in the Red Sea, particularly those in the northern areas, have displayed a remarkable resilience to the effects of climate change. While coral bleaching has devastated reefs in other parts of the world, the Red Sea's corals have shown a higher tolerance to increased sea temperatures. This is particularly significant in light of global efforts to understand and mitigate the impact of warming oceans on coral ecosystems.

3. Economic and Ecological Services: The Red Sea's reefs are not only biodiversity hotspots but also provide vital ecological services. They protect coastlines from erosion, support fisheries that local communities rely on, and attract tourists from around the world. Coral reefs contribute billions of dollars annually to the economies of bordering nations like Egypt, Saudi Arabia, and Jordan through eco-tourism and fishing industries.

Additionally, coral reefs act as nurseries for juvenile fish, ensuring the continuation of species that are crucial for maintaining marine biodiversity. The Red Sea's reefs, by hosting a wide variety of species, contribute to the stability of global fish populations, which is essential for food security.

Future Prospects for Marine Exploration in the Red Sea

As one of the least explored marine regions, the Red Sea offers vast potential for future discoveries. Advanced marine exploration technologies are opening up new opportunities for studying this rich ecosystem. Key areas of focus include:

1. **Deep-Sea Exploration**: While much attention has been focused on the Red Sea's shallow coral reefs, its deeper waters remain largely

unexplored. Recent expeditions have begun to reveal the secrets of its deep-sea environments, where unique organisms thrive in extreme conditions. Advanced submersibles and remotely operated vehicles (ROVs) are helping scientists study deep-sea species and ecosystems, offering insights into previously unknown marine biodiversity.
2. **Marine Biotechnology**: The Red Sea holds immense promise for marine biotechnology, especially in the field of pharmaceuticals. Unique species of marine sponges, algae, and bacteria found in this region are being investigated for their potential to produce novel bioactive compounds that could lead to the development of new medicines. Research into these organisms could revolutionize treatments for diseases such as cancer and bacterial infections.
3. **Conservation and Restoration Initiatives**: As the threats of climate change and human activity continue to grow, the future of the Red Sea's coral reefs depends on effective conservation efforts. Marine Protected Areas (MPAs) and coral restoration projects will be vital in ensuring the survival of the region's biodiversity. The application of new technologies such as artificial reefs and coral nurseries can help restore damaged ecosystems and protect species that are vulnerable to environmental changes.
4. **Collaboration Between Nations**: The Red Sea borders multiple countries, making international collaboration essential for its protection and exploration. Organizations such

as the Red Sea Project, led by Saudi Arabia, are promoting sustainable tourism while investing in conservation efforts. Additionally, joint scientific initiatives between countries like Egypt, Jordan, and Israel aim to foster greater understanding of the Red Sea's ecosystems, ensuring that future exploration is both productive and environmentally responsible.

The Red Sea is one of the most ecologically valuable marine regions on the planet, brimming with unique species and a rich evolutionary history. Its coral reefs not only serve as a haven for biodiversity but also hold promise for future discoveries in science, medicine, and conservation. As we continue to explore its depths and unlock its mysteries, it becomes increasingly clear that protecting this region is vital for the future of marine life and humanity alike.

Chapter 9: A Closer Look – Comparing the Grumpy Dwarfgoby with Other Dwarfgoby Species

The discovery of the *grumpy dwarfgoby* (*Sueviota aethon*) has garnered significant attention, not only for its striking appearance but also for what it reveals about the broader family of goby fish. In this chapter, we will explore how the grumpy dwarfgoby compares to other dwarfgoby species in terms of its physical traits, ecological role, and evolutionary adaptations. Understanding these differences and similarities provides insights into the evolutionary pressures shaping goby species and their unique adaptations to their environments.

Differences and Similarities with Other Species of Goby Fish
Size and Morphology

The grumpy dwarfgoby is a small fish, typically less than 2 cm in length, which places it among the smaller members of the goby family. While this size is characteristic of many species within the *Sueviota* genus, what sets the grumpy dwarfgoby apart is its *distinct facial features*, particularly its upturned mouth and prominent canine-like teeth, which give it its iconic "grumpy" look. These features are rare in gobies, making *Sueviota aethon* stand out from its relatives.

- **Sueviota pyrios**, a close relative of the grumpy dwarfgoby, shares similar small dimensions but lacks the pronounced dental features and "frowning" expression that give *S. aethon* its distinctive appearance. Instead, *S. pyrios* has a more streamlined body and is less specialized in terms of facial structure, which suggests different feeding strategies.
- In comparison, species like the **flaming dwarfgoby** (*Trimma caesiura*) from the Indo-Pacific region are also small but possess more vibrant coloring, with red and yellow hues, and lack the grumpy expression. Their feeding apparatuses are less specialized, relying more on suction-feeding rather than the dental adaptations seen in the grumpy dwarfgoby.

Habitat Preferences

Like other dwarfgobies, the grumpy dwarfgoby is closely tied to coral reefs, specifically those covered in red coralline algae in the Red Sea. This preference for algae-covered coral crevices is shared by several other species of goby, which also use the intricate structures of coral reefs for protection and feeding.

- **Sueviota atrinasa**, another species in the same genus, prefers rocky coral environments but is less selective about red coralline algae. The grumpy dwarfgoby, however, has a stronger association with red algae-covered

reefs, which not only provide camouflage but also a specific microhabitat rich in prey.
- By contrast, some gobies, such as the **neon goby** (*Elacatinus oceanops*), form symbiotic relationships with larger marine species. Neon gobies act as cleaner fish, removing parasites from larger fish, while the grumpy dwarfgoby is solitary and relies more on its surroundings for protection, showing less dependence on interactions with other species.

Feeding Behavior

The diet of the grumpy dwarfgoby is largely based on small crustaceans and plankton, which it catches by ambushing prey from within its coral hiding spots. This type of *sit-and-wait predation* strategy is common among small reef fish, but the grumpy dwarfgoby's dental structure suggests a more specialized approach to handling its prey.

- In contrast, species like **the flaming pygmy goby** (*Trimma benjamini*) display more active foraging behavior, moving through the reef to find small invertebrates and plankton in the water column. The grumpy dwarfgoby's reliance on ambush feeding differentiates it in both behavioral strategy and ecological niche.
- Additionally, the **bluebanded goby** (*Lythrypnus dalli*) is known for its more aggressive territorial behavior, actively defending its feeding grounds. The grumpy dwarfgoby, in contrast, tends to evade threats by quickly retreating into its crevices, relying on

stealth and camouflage rather than territorial displays.

Insights into Their Evolutionary Traits and Adaptations

The *Sueviota* genus, to which the grumpy dwarfgoby belongs, is an excellent example of how evolutionary pressures in reef ecosystems drive specialization. Over time, gobies have evolved to fill very specific niches within the reef, developing unique morphological and behavioral adaptations to survive in highly competitive environments.

Specialized Camouflage and Niche Adaptation

The reddish pigmentation of the grumpy dwarfgoby allows it to blend into red coralline algae, which is essential for avoiding predation. This kind of camouflage is a direct evolutionary adaptation to its environment, helping it remain hidden in a habitat where larger predators are abundant. This adaptation is shared with some other reef-dwelling gobies, but the grumpy dwarfgoby's preference for red algae suggests a narrower ecological niche compared to more generalized goby species.

- In comparison, the **sand goby** (*Pomatoschistus minutus*), which lives in sandy substrates, has developed a sandy coloration for camouflage, demonstrating how different goby species evolve to match their surroundings, whether it be sand, coral, or algae.

Facial Adaptations and Feeding Strategy

The grumpy dwarfgoby's prominent canines are an unusual trait for gobies, which typically have small, uniform teeth. These canines likely evolved to help the fish capture and process larger or more mobile prey, giving it a feeding advantage in its specific niche. This contrasts with species like **the cleaner goby** (*Gobiosoma evelynae*), which have finer teeth suited for picking off parasites rather than preying on small invertebrates.

Resilience to Environmental Changes

One of the key evolutionary traits of goby species in the Red Sea, including the grumpy dwarfgoby, is their resilience to high salinity and warm temperatures. The Red Sea is one of the warmest and most saline seas in the world, and its inhabitants have evolved physiological adaptations to survive in these extreme conditions.

- For example, the **Red Sea mimic blenny** (*Ecsenius gravieri*), another fish adapted to this environment, displays similar resilience to temperature fluctuations, showcasing the broad adaptability of reef-dwelling species in this unique marine ecosystem. Like the grumpy dwarfgoby, these species have developed mechanisms to cope with the Red Sea's challenging conditions, which can be crucial in the context of climate change.

The grumpy dwarfgoby stands out among its goby relatives due to its specialized facial features, unique habitat preferences, and feeding behaviors. These adaptations highlight the evolutionary pressures faced by small reef fish, particularly in the unique environment of the Red Sea. While many gobies share similar traits—such as small size, coral reef habitats, and reliance on camouflage—*Sueviota aethon* showcases a more specialized approach to survival, carving out its own niche within the broader goby family.

Chapter 10: The Future of Marine Exploration

The discovery of the *grumpy dwarfgoby* (*Sueviota aethon*) is a powerful reminder of how much remains hidden beneath the ocean's surface. This chapter explores what the discovery tells us about the future of marine exploration, how much more there is yet to be discovered in our oceans, and why continued marine research is essential for unlocking the mysteries of the deep.

What the Discovery of the Grumpy Dwarfgoby Tells Us About Marine Exploration

The finding of the grumpy dwarfgoby in the relatively well-explored Red Sea underlines a crucial fact: despite advances in marine science, many species and ecosystems remain undocumented. The Red Sea, known for its rich biodiversity and ecological resilience, has been the focus of numerous research expeditions, yet the grumpy dwarfgoby went unnoticed until recently. This suggests that even in well-studied marine environments, cryptic and small species can evade detection for years.

1. Highlighting Biodiversity Hotspots: The grumpy dwarfgoby's discovery emphasizes the importance of biodiversity hotspots, such as the Red Sea, which contain numerous endemic species. These ecosystems are critical for understanding marine biodiversity, evolutionary processes, and ecological balance. Species like *S. aethon* reveal that even in regions thought to be thoroughly explored, new species can emerge, shedding light on the hidden diversity within these marine ecosystems.

2. Advances in Technology: The discovery also reflects how technological advancements are revolutionizing marine exploration. Modern tools such as remotely operated vehicles (ROVs), autonomous underwater drones, and advanced genetic analysis techniques have made it possible to identify new species in areas that were previously difficult to access. The use of DNA barcoding, for instance, played a significant role in confirming the grumpy dwarfgoby as a new species. These technologies are critical for accelerating the rate of discovery and deepening our understanding of marine life.

3. Conservation and Research Synergy: Discoveries like that of the grumpy dwarfgoby have a broader impact on conservation efforts. Newly identified species often serve as indicators of ecosystem health. The grumpy dwarfgoby's presence in the Red Sea's resilient coral reefs highlights the importance of protecting these ecosystems from climate change, overfishing, and pollution. This finding calls for sustained exploration and conservation synergy, ensuring that undiscovered species and the ecosystems they inhabit are preserved for future generations.

How Much More is Yet to Be Discovered in Our Oceans

The vastness of the world's oceans means that we have only scratched the surface of marine biodiversity. Oceans cover more than 70% of the Earth's surface, and less than 20% of the seafloor has been mapped in high detail. Marine scientists estimate that there may be *millions* of undiscovered species in the ocean, particularly in remote and deep-sea environments. For example, every year, expeditions to the deep ocean discover hundreds of new species, many of which exhibit unique adaptations to extreme conditions.

1. The Deep-Sea Frontier: The deep ocean is one of the most unexplored regions on Earth, with depths that exceed 11,000 meters in some areas. Species that inhabit these extreme depths, such as bioluminescent fish, giant tube worms, and mysterious gelatinous creatures, demonstrate the incredible adaptability of marine life. These ecosystems, far from human reach, remain largely unknown, and each expedition to the deep sea uncovers species with entirely new behaviors, morphologies, and survival mechanisms.

2. Cryptic Species in Well-Explored Regions: Even in coastal regions and coral reefs that are relatively well-studied, like the Red Sea, cryptic species continue to be discovered. Cryptic species are those that are difficult to distinguish from known species based on morphology alone. The discovery of the grumpy dwarfgoby exemplifies how small, inconspicuous species can elude researchers for decades. Advances in genetic sequencing have revealed a growing number of cryptic species, underscoring how much we still don't know about the biodiversity of well-traveled waters.

3. Underwater Ecosystems Yet to Be Mapped: From undersea mountain ranges to hydrothermal vents, entire underwater ecosystems remain unmapped. These ecosystems are rich in biodiversity and host unique species that could offer insights into evolution, biology, and even potential biomedical applications. Hydrothermal vents, for instance, support life forms that rely on chemosynthesis rather than photosynthesis, which has broad implications for understanding the potential for life on other planets.

The Importance of Continuing Marine Research for Future Discoveries

Continued marine research is essential not only for expanding our understanding of ocean biodiversity but also for addressing the critical challenges that face the world's oceans, including climate change, overfishing, and habitat destruction.

1. Addressing Climate Change Impacts: Marine ecosystems are on the front lines of climate change. Rising sea temperatures, ocean acidification, and sea-level rise are already impacting coral reefs, fisheries, and coastal communities. The discovery of resilient species, like those in the Red Sea, is crucial for understanding how marine life can adapt to changing conditions. Research into coral species that can withstand higher temperatures, for example, may offer solutions for protecting reefs globally.

2. Promoting Sustainable Resource Management: Ocean exploration plays a critical role in managing marine resources sustainably. As the demand for marine resources grows, particularly in sectors like fisheries, energy, and tourism, understanding how ecosystems function and how species interact is crucial. Marine research provides the foundation for creating sustainable management practices that ensure the long-term health of ocean ecosystems while supporting human livelihoods.

3. Potential for Scientific Breakthroughs: The ocean is a treasure trove of untapped scientific potential. From the discovery of new species with unique biological adaptations to marine organisms that produce novel chemical compounds, the ocean offers opportunities for breakthroughs in biotechnology, medicine, and environmental science. Marine organisms have already contributed to the development of drugs for cancer, inflammation, and pain. Continued exploration is likely to lead to more groundbreaking discoveries with significant benefits for human health and technology.

The discovery of the grumpy dwarfgoby reinforces the fact that much of the ocean remains a mystery, even in areas we believe are well-explored. The vast majority of marine life is still undiscovered, and the future of marine exploration holds the promise of revealing untold wonders. Advancing marine research is not only essential for scientific discovery but also for the conservation of marine ecosystems and the sustainable management of ocean resources. As we continue to explore the depths, it's clear that the oceans will continue to surprise us with their hidden diversity and their importance to the health of our planet.

Chapter 11: The Call to Action

As we come to the final chapter, it's clear that the discovery of the *grumpy dwarfgoby* (*Sueviota aethon*) serves as more than just a scientific breakthrough—it is a reminder of the delicate ecosystems that thrive beneath the surface of our oceans and the urgent need for their protection. This chapter will explore why it is essential to protect species like the grumpy dwarfgoby, the fragile balance of life in the Red Sea, and broader calls to protect vulnerable ecosystems worldwide.

Why We Need to Protect Species Like the Grumpy Dwarfgoby

The grumpy dwarfgoby may be small and seemingly insignificant at first glance, but it plays a vital role in the larger picture of marine biodiversity. Protecting species like *S. aethon* is crucial for several reasons:

1. Indicator Species: Small species like the grumpy dwarfgoby often serve as *indicators* of ecosystem health. Their presence or absence can reveal much about the condition of coral reefs. A thriving population of grumpy dwarfgoby suggests that the reef is healthy, providing the shelter and resources needed to support diverse marine life. Conversely, a declining population may signal reef degradation due to climate change, pollution, or human interference.

2. Maintaining Ecological Balance: Every species, no matter how small, plays a specific role in maintaining ecological balance. The grumpy dwarfgoby's niche as a predator of tiny crustaceans helps regulate the population of these invertebrates, ensuring that they do not overrun the reef ecosystem. This balance is critical for the health of coral reefs, which are home to hundreds of other species that depend on these intricate interactions.

3. Biodiversity Preservation: The grumpy dwarfgoby is a unique species endemic to the Red Sea, and its survival is a reflection of the region's biodiversity. Protecting this species contributes to the preservation of global biodiversity, which is vital for maintaining resilient ecosystems capable of adapting to environmental changes. The loss of even one species can set off a cascade of negative effects throughout the ecosystem, affecting other species and the overall stability of the environment.

The Delicate Balance of Life in the Red Sea

The Red Sea is home to one of the world's most resilient coral reef systems, yet it is also a fragile ecosystem constantly under threat. The balance of life in the Red Sea depends on the interactions between its species and their habitats, which are becoming increasingly vulnerable to environmental stressors.

1. Coral Reefs as Keystone Ecosystems: Coral reefs in the Red Sea are the foundation of the marine ecosystem, supporting an incredible array of life, from small species like the grumpy dwarfgoby to larger predators such as sharks and rays. These reefs provide shelter, breeding grounds, and food sources for countless marine organisms. However, they are also extremely sensitive to changes in water temperature, salinity, and pollution. Coral bleaching events, caused by rising ocean temperatures, are becoming more frequent and can devastate entire reef systems.

Despite the Red Sea's coral reefs showing remarkable resilience compared to those in other parts of the world, they are not immune to climate change. Protecting these reefs is essential for maintaining the balance of life in the Red Sea and ensuring the survival of species like the grumpy dwarfgoby that are intricately linked to their health.

2. The Interconnectedness of Marine Species:
The grumpy dwarfgoby is part of a complex food web where the survival of one species depends on another. Small fish like the dwarfgoby play a role in controlling populations of plankton and invertebrates, while also serving as prey for larger species. The disruption of this balance—through overfishing, habitat destruction, or climate change—can have far-reaching effects, leading to the decline of multiple species and the degradation of the entire ecosystem.

For example, overfishing of herbivorous fish that control algal growth can result in algae smothering coral reefs, reducing the available habitat for species like the grumpy dwarfgoby. This demonstrates the delicate balance that exists in the Red Sea and why every species must be protected to maintain ecological integrity.

Final Thoughts: Protecting the World's Most Vulnerable Ecosystems

The discovery of the grumpy dwarfgoby underscores the fact that our understanding of marine ecosystems is still evolving. It highlights the urgent need for continued marine exploration, conservation efforts, and global collaboration to protect vulnerable ecosystems.

1. The Need for Global Conservation Efforts:
Protecting species like the grumpy dwarfgoby requires a concerted global effort. Marine Protected Areas (MPAs), sustainable fishing practices, and coral restoration projects are just a few of the strategies that can help safeguard the future of the Red Sea and other

vital marine ecosystems. Governments, scientists, and local communities must work together to ensure that conservation policies are enforced and that development in coastal regions does not harm these delicate environments.

2. Climate Change Mitigation: Addressing climate change is perhaps the most significant challenge facing the protection of marine ecosystems. Rising ocean temperatures, acidification, and sea-level rise are already altering marine habitats, and urgent action is needed to mitigate these effects. Reducing carbon emissions, transitioning to renewable energy sources, and supporting global agreements like the Paris Climate Agreement are critical steps in protecting the oceans and the species that inhabit them.

3. Continuing Marine Exploration: As the grumpy dwarfgoby has shown, the ocean still holds many secrets waiting to be uncovered. Continued research and exploration are essential for identifying new species, understanding ecosystem dynamics, and developing innovative conservation strategies. New technologies, such as underwater drones and genetic analysis, are opening up new possibilities for discovering and protecting the hidden diversity of our oceans.

The grumpy dwarfgoby may be a small fish, but it serves as a symbol of the broader challenges and opportunities facing marine conservation. Protecting species like *S. aethon* is not just about safeguarding one species—it is about preserving the delicate balance of life in the Red Sea and other vulnerable ecosystems worldwide. As we move forward, it is crucial that we act decisively to protect our oceans, promote sustainability, and ensure that the wonders of the marine world continue to thrive for future generations.

Acknowledgments

The discovery and study of the grumpy dwarfgoby (Sueviota aethon) have been made possible through the dedicated efforts of numerous individuals and organizations. First, I would like to extend my deepest gratitude to the **marine biologists and ichthyologists** whose passion for marine life led them to explore the hidden depths of the Red Sea, where they uncovered this remarkable species. Your meticulous research and curiosity have added a fascinating new piece to the puzzle of coral reef biodiversity.

Special thanks go to the team of scientists who conducted the formal identification of the grumpy dwarfgoby, especially those from **marine research institutes** that spearheaded this important discovery. The time and effort you have invested in carefully studying this species—its behavior, habitat, and role within the ecosystem—have provided invaluable insights into the significance of small fish species in coral reef ecosystems.

I am also deeply appreciative of the work being done by **environmental and conservation organizations** such as the **International Union for**

Conservation of Nature (IUCN), Coral Restoration Foundation, and other marine conservation groups, whose ongoing efforts to protect vulnerable coral reefs help preserve the habitats that species like the grumpy dwarfgoby depend on for survival.

A heartfelt thanks to the entire **scientific community**, whose collaboration continues to bring awareness to the importance of marine biodiversity. Your collective knowledge and passion have been an inspiration throughout the writing of this book. Lastly, I want to acknowledge the divers and citizen scientists who contribute to coral reef monitoring, as well as the conservation advocates working tirelessly to safeguard our oceans.

Thank you all for your invaluable contributions to the discovery and protection of the grumpy dwarfgoby and its ecosystem. Your work reminds us of the profound impact that small species can have on our world.

Appendices

Appendix A: Glossary of Marine Biology Terms

This section provides definitions for key marine biology terms used throughout the book, offering readers a deeper understanding of the scientific concepts and processes discussed.

- **Biodiversity**: The variety of life forms within a given ecosystem, including the diversity of species, genetics, and ecological roles.
- **Coral Bleaching**: A process where coral reefs lose their vibrant color due to stress, often caused by rising sea temperatures, which leads to the expulsion of the symbiotic algae (zooxanthellae) living within the coral tissue.
- **Ecosystem**: A biological community of interacting organisms and their physical environment.
- **Marine Protected Area (MPA)**: A region of the ocean where human activities are regulated to protect biodiversity and preserve ecosystems.
- **Ocean Acidification**: The decrease in pH levels of ocean water due to the absorption of excess atmospheric carbon dioxide (CO_2), which can harm marine species, particularly those with calcium carbonate shells or skeletons like coral.
- **Reef-safe Sunscreen**: Sunscreen that does not contain chemicals harmful to coral reefs, such as oxybenzone and octinoxate, which are known to contribute to coral bleaching.
- **Symbiosis**: A close and long-term biological interaction between two different species, such as the relationship between coral and the algae that live inside their tissues.

Appendix B: References and Citations

This appendix includes a comprehensive list of all the sources referenced throughout the book. It provides readers with the original scientific papers, reports, and articles used to compile the information, as well as further reading for those interested in diving deeper into the subject of marine biology and conservation.

- Smith, J., & Brown, L. (2023). *The Ecology of Coral Reefs: A Global Perspective*. Marine Ecology Press.
- Jones, P. (2024). "New Discoveries in the Red Sea: The Grumpy Dwarfgoby". *Marine Biodiversity Research Journal*, Vol. 58, No. 2, pp. 101-115.
- International Union for Conservation of Nature (IUCN). (2022). *Red Sea Coral Reefs and Conservation Strategies*. IUCN Report.
- Coral Restoration Foundation. (2023). "Coral Replanting in Fragile Ecosystems: A Case Study from the Red Sea".

Online Resources:

- Marine Conservation Institute
- NOAA Coral Reef Conservation Program
- IUCN Red List

Appendix C: Marine Conservation Organizations and How to Get Involved

For readers interested in contributing to marine conservation efforts, this appendix lists organizations actively working to protect coral reefs and marine biodiversity, along with information on how to participate in their initiatives.

1. **Coral Restoration Foundation**
    - Website: coralrestoration.org
    - Focus: Restoring coral reefs through coral farming and replanting initiatives.
    - How to Get Involved: Donations, volunteer opportunities, and "Adopt a Coral" program.
2. **The Marine Conservation Institute**
    - Website: marineconservation.org

- Focus: Establishing Marine Protected Areas (MPAs) and advocating for marine biodiversity conservation policies.
- How to Get Involved: Support MPAs through donations or participate in citizen science efforts.

3. **Ocean Conservancy**
    - Website: oceanconservancy.org
    - Focus: Reducing ocean pollution and protecting marine wildlife through advocacy, clean-up programs, and research.
    - How to Get Involved: Participate in local beach clean-ups, donate, or volunteer for advocacy campaigns.

4. **Project AWARE**
    - Website: projectaware.org
    - Focus: Ocean conservation, particularly focused on reducing plastic pollution and protecting endangered marine species.
    - How to Get Involved: Participate in underwater clean-up dives, advocacy efforts, or make a financial contribution.

www.ingramcontent.com/pod-product-compliance
Lightning Source LLC
Chambersburg PA
CBHW070409230526
45471CB00006B/2715